THE METHODS AND SCOPE
OF
GENETICS

THE METHODS AND SCOPE

OF

GENETICS

AN INAUGURAL LECTURE DELIVERED
23 OCTOBER 1908

by

W. BATESON, M.A., F.R.S.

PROFESSOR OF BIOLOGY IN THE UNIVERSITY OF CAMBRIDGE

CAMBRIDGE:

at the University Press

1908

CAMBRIDGE
UNIVERSITY PRESS

University Printing House, Cambridge CB2 8BS, United Kingdom

Published in the United States of America by Cambridge University Press, New York

Cambridge University Press is part of the University of Cambridge.

It furthers the University's mission by disseminating knowledge in the pursuit of
education, learning and research at the highest international levels of excellence.

www.cambridge.org
Information on this title: www.cambridge.org/9781107652583

© Cambridge University Press 1908

First published 1908
First paperback edition 2014

A catalogue record for this publication is available from the British Library

ISBN 978-1-107-65258-3 Paperback

PREFATORY NOTE

THE Professorship of Biology was founded in 1908 for a period of five years partly by the generosity of an anonymous benefactor, and partly by the University of Cambridge. The object of the endowment was the promotion of inquiries into the physiology of Heredity and Variation, a study now spoken of as Genetics.

It is now recognized that the progress of such inquiries will chiefly be accomplished by the application of experimental methods, especially those which Mendel's discovery has suggested. The purpose of this inaugural lecture is to describe the outlook over this field of research in a manner intelligible to students of other parts of knowledge.

W. B.

28 *October*, 1908

THE METHODS AND SCOPE OF GENETICS

THE opportunity of addressing fellow-students pursuing lines of inquiry other than his own falls seldom to a scientific man. One of these rare opportunities is offered by the constitution of the Professorship to which I have had the honour to be called. That Professorship, though bearing the comprehensive title "of Biology," is founded with the understanding that the holder shall apply himself to a particular class of physiological problems, the study of which is denoted by the term Genetics. The term is new; and though the problems are among the oldest which have vexed the human mind, the modes by which they may be successfully attacked are also of modern invention. There is therefore a certain fitness in the employment of this occasion for the deliverance of a discourse explaining something of the aims of Genetics and of the methods by which we trust they may be reached.

You will be aware that the claims put forward in the name of Genetics are high, but I trust to be able to show you that they are not high without reason. It is the ambition of every one who in youth devotes himself to the search for natural truth, that his work may be found somewhere in the main stream of progress. So long only as he keeps something of the limitless hope with which his voyage of discovery began, will his courage and his spirit last. The moment we most dread is one in which it may appear that, after all, our effort has been spent in exploring some petty tributary, or worse, a backwater of the great current. It is because Genetic research is still pushing forward in the central undifferentiated trunk of biological science that we confess no guilt of presumption in declaring boldly that whatever difficulty may be in store for those who cast in their lot with us, they need fear no disillusionment or misgiving that their labour has been wasted on a paltry quest.

In research, as in all business of exploration, the stirring times come when a fresh region is suddenly unlocked by the discovery of a new key. Then conquest is easy and there are prizes for all. We are happy in that during our own time not a few such territories have been revealed to the vision of mankind. I do not dare to suggest that in magnitude or splendour the field of Genetics may be compared with that now being disclosed to the physicist or the astronomer; for the glory of the celestial is one and the glory of the terrestrial is another. But I will say that for once to the man of ordinary power who cannot venture into those heights beyond, Mendel's clue has shown the way into a realm of nature which for surprising novelty and adventure is hardly to be excelled.

It is no hyperbolical figure that I use when I speak of Mendelian discovery leading us into a new world, the very existence of which was unsuspected before.

The road thither is simple and easy to follow. We start from a common fact, familiar to everyone, that all the ordinary animals and plants began their individual life by the union of two cells, the one male, the other female. Those cells are known as germ-cells or *gametes*, that is to say, "marrying" cells.

Now obviously the diversity of form which is characteristic of the animal and plant world must be somehow represented in the gametes, since it is they which bring into each organism all that it contains. I am aware that there is interplay between the organism and the circumstances in which it grows up, and that opportunity given may bring out a potentiality which without that opportunity must have lain dormant. But while noting parenthetically that this question of opportunity has an importance, which some day it may be convenient to estimate, the one certain fact is that all the powers, physical and mental that a living creature possesses were contributed by one or by both of the two germ-cells which united in fertilisation to give it existence. The fact that *two* cells are concerned in the production of all the ordinary forms of life was discovered a long while ago, and has been part of the common stock of elementary knowledge of all educated persons for about half a century.   The full consequences of this double nature seem nevertheless to have struck nobody before Mendel. Simple though the fact is, I have noticed that to many it is difficult to assimilate as a working idea.   We are accustomed to think of a man, a butterfly, or an apple tree as each *one* thing.   In order to understand the significance of Mendelism we must get thoroughly familiar with the fact that they are each *two* things, double throughout every part of their composition.   There is perhaps no better exercise as a preparation for genetic research than to examine the people one meets in daily life and to try in a rough way to analyse them into the two assemblages of characters which are united in them.   That we are assemblages or medleys of our parental characteristics is obvious.

We all know that a man may have his father's hair, his mother's colour, his father's voice, his mother's insensibility to music, and so on, but that is not enough.

Such an analysis is true, inasmuch as the various characters *are* transmitted independently, but it misses the essential point. For in each of these respects the individual is double; and so to get a true picture of the composition of the individual we have to think how *each* of the two original gametes was provided in the matter of height, hair, colour, mathematical ability, nail-shape, and the other features that go to make the man we know. The contribution of each gamete in each respect has thus to be separately brought to account. If we could make a list of all the ingredients that go to form a man and could set out how he is constituted in respect of each of them, it would not suffice to give one column of values for these ingredients, but we must rule two columns, one for the ovum and one for the spermatozoon, which united in fertilisation to form that man, and in each column we must represent how that gamete was supplied in respect of each of the ingredients in our list. When the problem of heredity is thus represented we can hardly avoid discovering, by mere inspection, one of the chief conclusions to which genetic research has led. For it is obvious that the contributions of the male and female gametes may in respect of any of the ingredients be either the same, or different. In any case in which the contribution made by the two cells is the same, the resulting organism—in our example the man—is, as we call it, *pure-bred* for that ingredient, and in all respects in which the contribution from the two sides of the parentage is dissimilar the resulting organism is *cross-bred*.

To give an intelligible account of the next step in the analysis without having recourse to precise and technical language is not very easy.

We have got to the point of view from which we see the individual made up of a large number of distinct ingredients, contributed from two sources, and in respect of any of them he may have received two similar portions or two dissimilar portions. We shall not go far wrong if we extend and elaborate our illustration thus. Let us imagine the contents of a gamete as a fluid made by taking a drop from each of a definite number of bottles in a chest, containing tinctures of the several ingredients. There is one such chest from which the male gamete is to be made up, and a similar chest containing a corresponding set of bottles out of which the components of the female gamete are to be taken.   But in either chest one or more of the bottles may be empty; then nothing goes in to represent that ingredient from that chest, and if corresponding bottles are empty in both chests, then the individual made on fertilisation by mixing the two collections of drops together does not contain the missing ingredient at all.   It follows therefore that an individual may thus be "pure-bred," namely alike on both sides of his composition as regards each ingredient in one of two ways, either by having received the ingredient from the male chest and from the female, or in having received it from neither.   Conversely in respect of any ingredient he may be "cross-bred," receiving the presence of it from one gamete and the absence of it from the other.

The second conception with which we have now to become thoroughly familiar is

that of the individual as composed of what
we call presences and absences of all the
possible ingredients. It is the basis of all
progress in genetic analysis. Let me give
you two illustrations. A blue eye is due to
the absence of a factor which forms pigment
on the front of the iris. Two blue-eyed
parents therefore, as Hurst has proved, do
not have dark-eyed children. The dark eye
is due to either a single or double dose of
the factor missing from the blue eye. So
dark-eyed persons may have families all dark-
eyed, or families composed of a mixture of
dark and light-eyed children in certain pro-
portions which on the average are definite.

Two plants of *Oenothera* which I exhibit
illustrate the same thing. One of them is
the ordinary *Lamarckiana*. I bend its stem.
It will not break, or only breaks with difficulty on account of the tough fibres it con-

tains. The stem of the other, one of de
Vries' famous mutations, snaps at once like
short pastry, because it does not contain the
factor for the formation of the fibres. Such
plants may be sister-plants produced by the
self-fertilisation of one parent, but they are
distinct in their composition and properties
—and this distinction turns on the presence
or absence of elements which are treated as
definite entities when the germ-cells are
formed. When we speak of such qualities as
the formation of pigment in an eye, or the
development of fibres in a stem, as due to
transmitted elements or factors, you will perhaps ask if we have formed any notion as to
the actual nature of those factors. For my
own part as regards that ulterior question I
confess to a disposition to hold my fancy on
a tight rein. It cannot be very long before
we shall *know* what some of the factors are,

and we may leave guessing till then. Meanwhile however there is no harm in admitting that several of them behave much as if they were ferments, and others as if they constructed the substances on which the ferments act. But we must not suppose for a moment that it is the ferment, or the objective substance, which is transmitted. The thing transmitted can only be the power or faculty to produce the ferment or the objective substance.

So far we have been considering the synthesis of the individual from ingredients brought into him by the two gametes. In the next step of our consideration we reverse the process, and examine how the ingredients of which he was originally compounded are distributed among the gametes that are eventually budded off from him.

Take first the case of the components in

respect of which he is pure-bred. Expectation would naturally suggest that all the germ-cells formed from him would be alike in respect of those ingredients, and observation shows, except in the rare cases of originating variations, the causation of which is still obscure, that this expectation is correct.

Hitherto though without experimental evidence no one could have been certain that the facts were as I have described them, yet there is nothing altogether contrary to common expectation. But when we proceed to ask how the germ-cells will be constituted in the case of an individual who is cross-bred in some respect, containing that is to say, an ingredient from the one side of his parentage and not from the other, the answer is entirely contrary to all the preconceptions which either science or common sense had formed about heredity. For we find definite experimental proof in nearly all the cases which have been examined, that the germ-cells formed by such individuals do either contain or not contain a representation of the ingredient, just as the original gametes did or did not contain it.

If *both* parent-gametes brought a certain quality in, then all the daughter gametes have it; if neither brought it in, then none of the daughter gametes have it. If it came in from one side and not from the other, then on an average in half the resulting gametes it will be present and from half it will be absent. This last phenomenon, which is called segregation, constitutes the essence of Mendel's discovery.

So recurring to the simile of the man as made by the mixing of tinctures, the process of redistribution of his characters among the germ-cells may be represented as a sorting

back of the tinctures again into a double row
of bottles, a pair corresponding to each ingredient; and each of the germ-cells as then
made of a drop from one or other bottle
of each pair: and in our model we may represent the phenomenon of segregation in a
crude way by supposing that the bottles
having no tincture in them, instead of being
empty contained an inoperative fluid, say
water, with which the tincture would not mix.
When the new germ-cells are formed, the two
fluids instead of diluting each other simply
separate again.   It is this fact which entitles
us to speak of the purity of germ-cells.   They
are pure in the possession of an ingredient,
or in not possessing it; and the ingredients,
or factors, as we generally call them, are units
because they are so treated in the process of
formation of the new gametes and because
they come out of the process of segregation

in the same condition as they went in at fertilisation.

As a consequence of these facts it follows that however complex may be the origin of two given parents the composition of the off-spring they can produce is limited. There is only a limited number of types to be made by the possible recombinations of the parental ingredients, and the relative numbers in which each type will be represented are often pre-dicable by very simple arithmetical rules.

For example, if neither parent possesses a certain factor at all, then none of the off-spring will have it. If either parent has two doses of the factor then all the children will have it; and if either parent has one dose of the factor and the other has none, then on an average half the family will have it, and half be without it.

To know whether the parent possesses

the factor or not may be difficult for reasons
which will presently appear, but often it is
quite easy and can be told at once, for there
are many factors which cannot be present in
the individual without manifesting their presence. I may illustrate the descent of such
a factor by the case of a family possessing
a peculiar form of night-blindness. The
affected individuals marrying with those
unaffected have a mixture of affected
and unaffected children, but their unaffected
children not having the responsible ingredient cannot pass it on[1].

[1] The investigation of this remarkable family was made
originally by Cunier. The facts have been reexamined and the
pedigree much extended by Nettleship. The numerical results
are somewhat irregular, but it is especially interesting as being
the largest pedigree of human disease or defect yet made. It
contains 2121 persons, extending over ten generations. Of these
persons, 135 are known to have been night-blind. In no single
case was the peculiarity transmitted through an unaffected
member. It should be mentioned that for night-blindness such
a system of descent is peculiar. More usually it follows the
scheme described for colour-blindness. It is not known wherein
the peculiarity of this family consists.

In such an observation two things are strikingly exemplified, (1) the fact of the permanence of the unit, and (2) the fact that a *mixture* of types in the family means that one or other parent is cross-bred in some respect, and is giving off gametes of more than one type.

The problem of heredity is thus a problem primary analytical. We have to detect and enumerate the factors out of which the bodies of animals and plants are built up, and the laws of their distribution among the germ-cells. All the processes of which I have spoken are accomplished by means of cell-divisions, and in the one cell-union which occurs in fertilisation. If we could watch the factors segregating from each other in cell-division, or even if by microscopic examination we could recognize this multitudinous diversity of composition that must certainly exist among the germ-cells of all ordinary individuals, the work of genetics would be much simpler than it is.

But so far no such direct method of observation has been discovered. In default we are obliged to examine the constitution of the germ-cells by experimental breeding, so contrived that each mating shall test the composition of an individual in one or more chosen respects, and, so to speak, sample its germ-cells by counting the number of each kind of offspring which it can produce. But cumbersome as this method must necessarily be, it enables us to put questions to Nature which never have been put before. She, it has been said, is an unwilling witness. Our questions must be shaped in such a way that the only possible answer is a direct "Yes" or a direct "No." By putting such questions we have received some astonishing answers which go far below the surface. Amazing though they be, they are nevertheless true; for though our witness may prevaricate, she cannot lie. Piecing these answers together, getting one hint from this experiment, and another from that, we begin little by little to reconstruct what is going on in that hidden world of gametes. As we proceed, like our brethren in other sciences, we sometimes receive answers which seem inconsistent or even contradictory. But by degrees a sufficient body of evidence can be attained to show what is the rule and what the exception. My purpose today must be to speak rather of the regular than of the irregular.

One clear exception I may mention. Castle finds that in a cross between the long-eared lop-rabbit and a short-eared breed, ears of intermediate length are produced: and that these intermediates breed approximately true.

Exceptions in general must be discussed elsewhere. Nevertheless if I may throw out a word of counsel to beginners, it is : Treasure your exceptions! When there are none, the work gets so dull that no one cares to carry it further. Keep them always uncovered and in sight. Exceptions are like the rough brickwork of a growing building which tells that there is more to come and shows where the next construction is to be.

You will readily understand that the presentation here given of the phenomena is only the barest possible outline. Some of the details we may now fill in. For example, I have spoken of the characters of the organism, its colour, shape, and the like, as if they were due each to one ingredient or factor. Some of them are no doubt correctly so represented; but already we know numerous bodily features which need the concurrence of several factors to produce them. Never-theless though the character only appears when all the complementary ingredients are together present, each of these severally and independently follows, as regards its trans-mission, the simple rules I have described.

This complementary action may be illus-trated by some curious results that Mr Punnett and I have encountered when ex-perimenting with the height of Sweet Peas. There are two dwarf varieties, one the prostrate "Cupid," the other the half-dwarf or "Bush" Sweet Peas. Crossed together they give a cross-bred of full height. There is thus some element in the Cupid which when it meets the complementary element from the Bush, produces the characteristic length of the ordinary Sweet Pea. We may note in passing that such a fact demonstrates at once the nature of Variation and Reversion. The Reversion occurs because the two factors that made the *height* of the old Sweet Pea again come together after being parted: and the Variations by which each of the dwarfs came into existence must have taken place by the dropping out of one of these elements or of the other.

Conversely there are factors which by their presence can prevent or inhibit the development and appearance of others present and unperceived.

For example, all the factors for pigmentation may be present in a plant or an animal; but in addition there may be another factor present which keeps the individual white, or nearly so.

There are cases in which the action of the factors is superposed one on top of the other, and not until each factor is removed in turn can the effects of the underlying factors be perceived. So in the mouse if no other colour-factor is present, the fur is chocolate. If the next factor in the series be there, it is black. If still another factor be added, it has the brownish grey of the common wild mouse. Conversely, by the variation which dropped out the top factor, a black mouse came into existence. By the loss of the black factor, the chocolate mouse was created, and for aught we can tell there may be still more possibilities hidden beneath.

In the disentanglement of the properties and interactions of these elementary factors, the science we must call to our aid is Physiological Chemistry. The relations of Genetics with the other branches of biology are close. Such work can only be conducted by those who have the good fortune to be able to count upon continual help and advice from specialists in the various branches of Zoology, Physiology, and Botany. Often we have questions with which only a cytologist can deal, and often it is the experience of a systematist we must invoke. The school of Genetics in Cambridge starts under happy auspices in that we are surrounded by colleagues qualified, and as we have often found, willing to give us such aid unstinted. But with chemical physiology, we stand in an even closer relation; and from the little I have dared to say respecting the action and interaction of factors, it is evident that for their disentanglement there must one day be an intimate and enduring partnership arranged with the physiological chemists.

Now, as the whole of the elaborate process by which the various elements are apportioned among the gametes must be got through in a few cell-divisions at most, and perhaps in one division only, it is not surprising that there is sometimes an interaction between factors that have quite distinct rôles to perform. These interactions are probably of several kinds. One, which I shall illustrate presently, is probably to be represented as a repulsion between two factors. As a consequence of its operations when the various factors are sorted out into the gametes, if the individual be cross-bred in respect of the *two* repelling factors, having received so to speak only a single dose of each, then the gametes are made up in such a way that each takes one or other of the two repelling factors, not both.

Mutual repulsions of this kind probably play a significant part in the phenomena of heredity. A single concrete case which Mr Punnett and I have been investigating for some years will illustrate several of these principles. We crossed together a pure white Sweet Pea having an erect standard, with another pure white Sweet Pea having a hooded standard. The result is, as you see, a purple flower with an erect standard. The colour comes from the concurrence of complementary elements. A dose of a certain ingredient from one parent meets a dose of another ingredient from the other parent and the two make pigment in the flower. From other experiments we know that the *purple* colour of the pigment is due to a dose of a third ingredient brought in from the hooded parent; and that in the absence of that blue factor, as we may call it, the flower would be red. The standard is erect because it contains a dose of the erectness-factor from the erect parent, and the hooded parent can readily be proved to owe its peculiar shape to the absence of that element.

Our purple plant is thus cross-bred for four factors, containing only one dose of each.

We let it fertilise itself, and its offspring show all the possible combinations of the four different factors and their absences which the genetic constitution of the plant can make.

Note that one of the combinations we expect to find is missing. There are white erect and white hooded—white because they are lacking one or other of the complementary ingredients necessary to the production of pigment. There are purple erect and purple hooded, of which the purple erect must perforce contain all the four factors, and the purple hooded must similarly contain all of them except that for erectness. But when we turn to the red class we are surprised to find that they are all erect, none hooded. One of the possible combinations is missing. If you examine this series of facts you will find there is only one possible interpretation: namely that the ingredient which turns the flower purple—alkalinity, perhaps we may call it—never goes into the same germ-cell as the ingredient which makes the standard erect. There are plenty of ways of testing the truth of this interpretation. For example, it follows that the purple erects from such a family will in perpetuity have offspring 1 purple hooded : 2 purple erect : 1 red erect; also that all the white hooded crossed with pure reds will give purples, and so on. These experiments have been made and the result has in each case been conformable to expectation.

Between these two factors, the purpleness and the erectness of standard, some antagonism or repulsion must exist. In some way therefore the chemical and the geometrical phenomena of heredity must be inter-related.

Some one will say perhaps this is all very well as a scientific curiosity, but it has nothing to do with real life. The right answer to such criticism is of course the lofty one that science and its applications are distinct: that the investigator fixes his gaze solely on the search for truth and that his attention must not be distracted by trivialities of application. But while we make this answer and at least try to work in the spirit it proclaims, we know in our hearts that it is a counsel of perfection. I suspect that even the astronomer who at his spectroscope is analysing the composition of Vega or Capella has still an eye sometimes free for the affairs of this planet, and at least the fact that his discoveries may throw light on our destinies does not diminish his zeal in their pursuit. And surely to the study of Heredity, preeminently among all the sciences, we are looking for light on human destiny. To pretend otherwise would be mere hypocrisy. So while reserving the higher line of defence I will reply that again and again in our experimental work we come very near indeed to human affairs. Sometimes this is obvious enough. No practical dog-breeder or seedsman can see the results of Mendelian recombination without perceiving that here is a bit of knowledge he can immediately apply. No sociologist can examine the pedigrees illustrating the simple descent of a deformity or a congenital disease, and not see that the new knowledge gives a solid basis for practical action by which the composition of a race could be modified if society so chose. More than this: we know for certain in one case, from the work of Professor Biffen, that the power to resist a disease caused by the invasion of a pathogenic organism, wheat-rust, is due to the presence of one of the simple factors or ingredients of which I have spoken, and what we know to be true in that one case we are beginning to suspect to be true of resistance to certain other diseases. No pathologist can see such an experiment as this of Professor Biffen's without realizing that here is a contribution of the first importance to the physiology of disease.

There is no lack of utility and direct application in the study of Genetics. I have alluded to some strictly practical results. If we want to raise mangels that will not run to seed, or to breed a cow that will give more milk in less time, or milk with more butter and less water, we can turn to Genetics with every hope that something can be done in these laudable directions. But here I would plead what I cannot but regard as a higher

usefulness in our work. Genetic inquiry aims
at providing knowledge that may bring, and
I think will bring, certainty into a region
of human affairs and concepts which might
have been supposed reserved for ages to be
the domain of the visionary. We have long
known that it was believed by some that our
powers and conduct were dependent on our
physical composition, and that other schools
have maintained that nurture not nature, to
use Galton's antithesis, had a preponderating
influence on our careers; but so soon as it
becomes common knowledge—not a philo-
sophical speculation, but a certainty—that
liability to a disease, or the power of resisting
its attack, addiction to a particular vice, or to
superstition, is due to the presence or absence
of a specific ingredient; and finally that these
characteristics are transmitted to the off-
spring according to definite, predicable rules,

then man's views of his own nature, his conceptions of justice, in short his whole outlook on the world, must be profoundly changed. Yet as regards the more tangible of these physical and mental characteristics there can be little doubt that before many years have passed the laws of their transmission will be expressible in simple formulae.

The blundering cruelty we call criminal justice will stand forth divested of natural sanction, a relic of the ferocious inventions of the savage. Well may such justice be portrayed as blind. Who shall say whether it is crime or punishment which has wrought the greater suffering in the world ? We may live to know that to the keen satirical vision of Sam Butler on the pleasant mountains of Erewhon there was revealed a dispensation, not kinder only, but wiser than the terrific code which Moses delivered from the flames of Sinai.

If there are societies which refuse to apply the new knowledge, the fault will not lie with Genetics. I think it needs but little observation of the newer civilisations to foresee that *they* will apply every scrap of scientific knowledge which can help them, or seems to help them in the struggle, and I am good enough Selectionist to know that in that day the fate of the recalcitrant communities is sealed.

The thrill of discovery is not dulled by a suspicion that the discovery can be applied. No harm is done to the investigator if he can resist the temptation to deviate from his aim. With rarest exceptions the discoveries which have formed the basis of physical progress have been made without any thought but for the gratification of curiosity. Of this there can be few examples more conspicuous than that which Mendel's work presents. Untroubled by any itch to make potatoes larger or bread cheaper, he set himself in the quiet of a cloister garden to find out the laws of hybridity, and so struck a mine of truth, inexhaustible in brilliancy and profit.

I will now suggest to you that it is by no means unlikely that even in an inquiry so remote as that which I just described in the case of the Sweet Pea, we may have the clue to a mystery which concerns us all in the closest possible way. I mean the problem of the physiological nature of Sex. In speaking of the interpretation of sexual difference suggested by our experimental work as of some practical moment, I do not imply that as in the other instances I have given, the knowledge is likely to be of immediate use to our species; but only that if true it makes a contribution to the stock of human ideas which no one can regard as insignificant.

In the light of Mendelian knowledge, when a family consists of more than one type the fact means that the germ-cells of one or other parent must certainly be of more than one kind. In the case of sex the members of the family are thus of two kinds, and the presumption is overwhelming that this distinction is due to a difference among the germ-cells. Next, since for all practical purposes the numbers of the two sexes produced are approximately equal, sex exhibits the special case in which a family consists of two types represented in equal numbers, half being male, half female. But I called your attention to the fact that equality of types results when *one* parent was cross-bred in the character concerned, having received one dose only of the factor on which it depends. So we may feel fairly sure that the distinction between the sexes depends on the presence in one or other of them of an unpaired factor. This conclusion appears to me to follow so immediately on all that we have learnt of genetic physiology that with every confidence we may accept it as representing the actual fact.

The question which of the two sexes contains the unpaired factor is less easy to answer, but there are several converging lines of evidence which point to the deduction that in Vertebrates at least, and in some other types, it is the female, and I feel little doubt that we shall succeed in proving that in them femaleness is a definite Mendelian factor absent from the male and following the ordinary Mendelian rules.

Before showing you how the Sweet Pea phenomenon aids in this inquiry I must tell you of some other experimental results. The first concerns the common currant moth, *Abraxas grossulariata*. It has a definite

pale variety called *lacticolor*. With these
two forms Doncaster has made a remarkable
series of experiments. When he began, *lacticolor* was only known as a female form. This
was crossed with the *grossulariata* male and
gave *grossulariata* only, showing that the
male was pure to type. The hybrids bred
together gave *grossulariata* males and females
and *lacticolor* females only. But the hybrid
males bred to *lacticolor* females produced
all four combinations, *grossulariata* males
and females, and *lacticolor* males and females.
When the *lacticolor* males were bred to *grossulariata* females, whether hybrid, or wild
from a district where *lacticolor* does not
exist, the result was that all the males were
grossulariata and all the females *lacticolor*!
It is difficult to follow the course of such an
experiment on once hearing and all I ask
you to remember is first that there is a series

of matings giving very curious distributions of the characters of type and variety among the two sexes. And then, what is perhaps the most singular fact of all, that the wild typical *grossulariata* female can when crossed with the *lacticolor* male produce all females *lacticolor*. This last fact can, we know, mean only one thing, namely that these wild females are in reality hybrids of *lacticolor*; though since the males are pure *grossulariata*, that fact would in the natural course of things never be revealed.

When we encounter such a series of phenomena as this, our business is to find a means of symbolical expression which will represent all the factors involved, and show how each behaves in descent. Such a system or scheme we have at length discovered, and I incline to think that it must be the true one. If you study this case you will find that there are nine distinct kinds of matings that can be made between the variety, the type and the hybrid, and the scheme fits the whole group of results. It is based on two suppositions:

1. That the female is cross-bred, or as we call it heterozygous for femaleness-factor, the male being without that factor. The eggs are thus each destined from the first to become either males or females, but as regards sex the spermatozoa are alike in being non-female.

2. That there is a repulsion between the femaleness-factor and the *grossulariata* factor.

Such a repulsion between two factors we are justified in regarding as possible because we have had proof of the occurrence of a similar repulsion in the case of the two factors in the Sweet Pea.

If the case of this moth stood alone it

would be interesting, but its importance is greatly increased by the fact that we know two cases in birds which are closely comparable. The simpler case to which alone I shall refer has been observed in the Canary. Like the Currant moth it has a kind of albino, called Cinnamon, and males of this variety when mated with ordinary dark green hen canaries produce dark males and Cinnamons which are always hens; while the green male and the Cinnamon hen produce nothing but greens of both sexes. This case, which has been experimentally studied by Miss Durham, offers a certain complication, but in its main outlines it is exactly like that of the moth, and the same interpretation is applicable to both.

The particular interpretation may be imperfect and even partially wrong; but that we are at last able to form a working idea of the course of such phenomena at all is a most encouraging fact. If we are right, as I am strongly inclined to believe, we get a glimpse of the significance of the popular idea that in certain respects daughters are apt to resemble their fathers and sons their mothers; a phenomenon which is certainly sometimes to be observed.

There are several collateral indications that we are on the right track in our theory of the nature of sex. One of these, derived from the peculiar inheritance of colour-blindness, is especially interesting. That affection, is common in men, rare in women. Men who are colour-blind can transmit the affection, but men who have normal vision cannot. Women however who are ostensibly normal may have colour-blind sons; and women who are colour-blind have, so far as we know, no sons who are not colour-blind[1].

[1] We have knowledge now of seven colour-blind women, having, in all, 17 sons who are all colour-blind. Most of these cases have been collected by Mr Nettleship.

Mendelian analysis of these facts shows that colour-blindness is due, not, as might have been supposed, to the absence of something from the composition of the body, but to the presence of something which affects the sight. Just as nicotine-poisoning can paralyse the colour sense, so may we conceive the development of a secretion in the body which has a similar action. The comparative exemption of the woman must therefore mean that there is in her a positive factor which counteracts the colour-blindness factor, and it is not improbable that the counteracting element is no other than the femaleness-factor itself.

I think I have said enough to prove that after all, those curiosities collected from observation of Sweet Peas and Canaries have no remote bearing on some very fascinating problems of human life.

Lastly I suppose it is self-evident that they have a bearing on the problem of Evolution. The facts of heredity and variation are the materials out of which all theories of Evolution are constructed. At last by genetic methods we are beginning to obtain such facts of unimpeachable quality, and free from the flaws that were inevitable in older collections. From a survey of these materials we see something of the changes which will have to be made in the orthodox edifice to admit of their incorporation, but he must be rash indeed who would now attempt a comprehensive reconstruction. The results of genetic research are so bewilderingly novel that we need time and an exhaustive study of their inter-relations before we can hope to see them in proper value and perspective. In all the discussions of the stability and fitness of species who ever contemplated the possibility of a wild species having one of its sexes permanently hybrid? When I spoke of adventures to be encountered in genetic research I was thinking of such astonishing discoveries as that.

There are others no less disconcerting. Who would have supposed it possible that the pollen-cells of a plant could be all of one type, and its egg-cells of two types? Yet Miss Saunders' experiments have provided definite proof that this is the condition of certain Stocks, of which the pollen grains all bear doubleness, while the egg-cells are some singles and some doubles. We cannot think yet of interpreting these complex phenomena in terms of a common plan. All that we know is that there is now open for our scrutiny a world of varied, orderly and specific physiological wonders into which we have as yet only peeped. To lay down positive propositions as to the origin and inter-relation of species in general, now, would be a task as fruitless as that of a chemist must have been who had tried to state the relationship of the elements before their properties had been investigated.

For the first time *Variation* and *Reversion* have a concrete, palpable meaning. Hitherto they have stood by in all evolutionary debates, convenient genii, ready to perform as little or as much as might be desired by the conjuror. That vaporous stage of their existence is over; and we see Variation shaping itself as a definite, physiological event, the addition or omission of one or more definite elements; and Reversion as that particular addition or subtraction which brings the total of the elements back to something it had been before in the history of the race.

The time for discussion of Evolution as

a problem at large is closed. We face that problem now as one soluble by minute, critical analysis. Lord Acton in his inaugural lecture said that in the study of history we are at the beginning of the documentary age. No one will charge me with disrespect to the great name we commemorate this year, if I apply those words to the history of Evolution: Darwin, it was, who first showed us that the species have a history that can be read at all. If in the new reading of that history, there be found departures from the text laid down in his first recension, it is not to his fearless spirit that they will bring dismay.